Routledge Revivals

Essays in Economic Theory

Essays in Economic Theory, first published in 1983, combines two essays on game theory and its applications in economics. The first, "Learning Behavior and the Noncooperative Equilibrium", considers whether an adaptive justification, like those commonly available for the optimization models frequently employed elsewhere in economics, can be found for the Nash noncooperative equilibrium. The second essay, "A Game of Fair Division", was motivated by the desire to find attractive methods for solving allocation problems and bargaining disputes that are simple enough to provide useful alternatives to existing methods. It studies in detail one such simple method: the classical "divide-and-choose" procedure. This book will be of interest to students of economics.

Essays in Economic Theory

Vincent P. Crawford

Routledge
Taylor & Francis Group

First published in 1983
by Garland Publishing, Inc.

This edition first published in 2015 by Routledge
2 Park Square, Milton Park, Abingdon, Oxon, OX14 4RN
and by Routledge
711 Third Avenue, New York, NY 10017

Routledge is an imprint of the Taylor & Francis Group, an informa business

Publisher's Note
The publisher has gone to great lengths to ensure the quality of this reprint but points out that some imperfections in the original copies may be apparent.

Disclaimer
The publisher has made every effort to trace copyright holders and welcomes correspondence from those they have been unable to contact.

A Library of Congress record exists under LC control number: 80008621

ISBN 13: 978-1-138-85862-6 (hbk)
ISBN 13: 978-1-315-71779-1 (ebk)

Essays in Economic Theory

Vincent P. Crawford

Garland Publishing, Inc.
New York & London, 1983

Library of Congress Cataloging in Publication Data

Crawford, Vincent P., 1950–
Essays in economic theory.

(Outstanding dissertations in economics)
Originally presented as the author's thesis (Ph.D.—MIT, 1976) With new pref.
Includes bibliographies.
1. Game theory—Addresses, essays, lectures.
I. Title. II. Series.
HB144.C72 1983 330.1 80-8621
ISBN 0-8240-4178-X

All volumes in this series are printed on acid-free, 250-year-life paper.

Printed in the United States of America

ACKNOWLEDGEMENTS

I owe my greatest debt to the members of my thesis committee -- Franklin Fisher, Hal Varian, and Martin Weitzman. Each contributed greatly to my interest in and knowledge of economic theory, as well as to the form and content of this thesis. Many others have made useful comments on earlier drafts of one or both essays, notably Robert Bishop, Jacques Crémer, Clifford Donn, Jerry Hausman, Marc Lonoff, James MacKinnon, Paul Samuelson, Robert Solow, and several anonymous referees. The manuscript was cheerfully deciphered and beautifully typed by Vicki Elms. Finally, she and several other friends -- Peter Berck, Bev Cayford, Bill Cheney, Sally Ann Dodge, Lex Kelso, Rick Mishkin, Jeff Perloff, Mary Kay Plantes, Bernie Reddy, Don Sillers, Alix Werth, and Steve Wiggins -- provided much encouragement and helped to create an atmosphere in which writing a thesis was not only tolerable, but a pleasure.

TABLE OF CONTENTS

TO MARJORIE AND BENNETT

PREFACE

Essays in Economic Theory combines two essays on game theory nd its applications in economics. The first, "Learning Behavior nd the Noncooperative Equilibrium," considers whether an adaptive ustification, like those commonly available for the optimization odels frequently employed elsewhere in economics, can be found for he Nash noncooperative equilibrium.[1] While it is known that such justification is often available when the equilibrium involves nly pure (or unrandomized) strategies, it is shown in this essay hat all members of a wide class of behaviorally plausible learning echanisms must fail to converge in "almost all" games in which the quilibrium involves mixed (or randomized) strategies.

An exact formal analogy is then developed, which allows hese learning mechanisms to be viewed as discrete tâtonnement pro- esses in properly chosen competitive exchange economies. The nstability of the mechanisms is elucidated by noting that in the nalogy, the functions that correspond to excess demand functions are, xcept at the boundaries, independent of the mixed-strategy prob- bilities that correspond to "own" prices: in "almost all" ompetitive economies with this property, the tâtonnement process is ecessarily unstable.

This instability does not necessarily extend to pure- trategy equilibria, because the independence noted above, which arises from the linearity in probabilities of von Neumann-Morgenstern utilities, does not hold in general in the pure-strategy case. Thu the sophistication required of agents for the mixed-strategy Nash equilibrium to be descriptively applicable is likely to be significantly greater than that required for the applicability of the pure-strategy equilibrium, or of non-interactive optimization models. This is so in spite of the fact that mixed strategies have long bee considered a successful generalization of pure strategies from a normative standpoint, and for the purpose of guaranteeing the existence of equilibrium under weak assumptions on the payoff functi and strategy spaces.

While the results in "Learning Behavior and the Non-cooperative Equilibrium" are unchanged from when they originally app in my thesis, several minor errors and confusions have been correcte In addition, the introduction and conclusions have been substantiall revised to reflect both subsequent changes in my thinking about the noncooperative equilibrium and developments in the literature that occurred, or of which I became aware, after 1976. I shall, therefor make no further attempt to put this essay in perspective here; it speaks for itself.

The second essay, "A Game of Fair Division," was motivate by the desire to find attractive methods for solving allocation problems and bargaining disputes that are simple enough to provide useful alternatives to existing methods.[2] It studies in detail one such simple method: the classical "divide-and-choose" procedure.

In a pure-exchange, multi-good setting, the noncooperative equilibrium of the game that arises when two agents who know each other's preferences agree to use the method is characterized and shown, under very general assumptions, to have several interesting properties. While the allocation of goods resulting from the game need not be Pareto-efficient in general, it is <u>fair</u> in the sense that neither agent envies the other's consumption; and it is an efficient point in the set of all fair allocations. The game confers an advantage on the agent who assumes the role of divider, and as a result its allocations treat agents more unequally than equal-income competitive allocations. When both agents have identical preferences, however, the Pareto-inefficiency, divider's advantage, and unequal treatment all disappear, and these results are robust in the sense that if agents have nearly identical preferences, nearly equal treatment and near-Pareto-efficiency result.

All of these results, except the one establishing the relationship between game allocations and equal-income competitive allocations, continue to hold when production is possible. And when the divide-and-choose process is followed by cooperative trade that players anticipate in their strategy choices, some of the results, notably the existence of divider's advantage, remain valid. However, none of the strong results obtained in the two-person case generalize to n-person versions of the divide-and-choose method.

In [3], I suggested that these results provide a theoretical basis for experimenting with the divide-and-choose method as a technique for resolving bargaining disputes in which each person is judged to have, in some sense, equal rights to the bundle of goods being allocated. Since then, much of my research has dealt with mechanisms related either directly or indirectly to the divide-and-choose method. In [9], many of the results in [3] are generalize to the case where some of the commodities in the bundle to be allocated are indivisible: as long as there are sufficient quantitie of divisible goods in the bundle as well, in a sense made precise in the paper, most of the results mentioned above remain valid.[3]

In [6], a procedure obtained by a simple modification of t divide-and-choose procedure, with almost uniformly superior optimalit properties, is developed and compared to the classical procedure. Th new procedure, which allows the agent whose role is analogous to the divider's to offer the other agent a choice between equal division (or, more generally, some other <u>basis allocation</u>) and an allocation t "divider" is free to specify, is equally simple and easy to use, fair and fully Paréto-efficient when agents are well informed about each other's preferences. It still confers an advantage on the divider, but this unequal treatment is uniformly less severe than in the classical method. In addition, its optimality properties are generally more robust. Thus, to the extent that [3] makes a case for experimenting with the divide-and-choose method, [6] makes an even stronger case for experimenting with the improved version developed there.

The procedure discussed in [6] appeared almost simultaneously, in slightly different forms, in several places in the theoretical literature on allocation mechanisms. Arrow [1] discusses its performance in a public-goods setting when agents do not have complete information about each other's preferences. Maskin [12] shows that an n-person generalization of it provides the only way to implement the social choice correspondence that selects individually rational and Pareto-efficient allocations, when the solution concept used to define implementation is Aumann's strong Nash equilibrium. Wilson [15] studies a different n-person procedure, which specializes to the one discussed in [6] when there are only two persons. He obtains results in some respects more general, and in others less general, than some of those obtained in [6]. Finally, I showed in [4] that if an n-person generalization of the procedure in [6] is modified by auctioning off the role of divider in a properly specified way, one obtains a procedure for generating Pazner and Schmeidler's [13] Pareto-efficient egalitarian-equivalent allocations; this procedure also provides a new, constructive proof of Pazner and Schmeidler's result establishing the existence of such allocations.

Another strand of research that began, for me, with the study of the divide-and-choose method is concerned with the compulsory-arbitration schemes that are now frequently used to resolve labor disputes in the public sector. When bargainers are well informed, the procedure in [6] is essentially equivalent to a modification of "final-offer" arbitration that I have called "multiple" final-offer arbitration. This procedure, which was originally suggested by Donn [10], is discussed extensively in [5], and is closely related to a procedure now in use in Eugene, Oregon. In [8], a formalization of some of Donn's ideas about this procedure is attempted under the polar assumption of extreme risk aversion. My recent survey [7] of the theoretical literature on compulsory arbitration provides additional references.

"A Game of Fair Division" appears substantially as it did originally in my thesis, except for the following corrections and substitutions:

(1) The proof of Theorem 2.4 incorporates a simplification suggested by a referee when [3] was under review.

(2) The original proof of Theorem 2.6 relied on the facts that, under the maintained assumptions, the utility-possibility frontier associated with the allocation problem is continuous and strictly downward-sloping and, when the divider and the chooser have identical utility functions, the feasible region of the divider's problem is mapped into the area in utility space below the 45° line and inside the frontier. This proof has been replaced by an argument I constructed to establish the analogous result, Theorem 2.4, in [9]. In addition to being mathematically more

lementary, the new argument has the advantage that, because it oes not depend as heavily on the structure of the utility-ossibility set, which is complex when indivisible commodities re present, it remains valid then.

(3) A lacuna in the proof of Theorem 2.7 has been illed.

(4) The original proof of Theorem 2.8,which was rroneous, has been replaced by a much shorter, correct proof.

Two postscripts should be added to the above. First, after A Game of Fair Division" was originally completed, reading [1] alerted e to the fact that the solution concept employed there is, technically, ot the ordinary Nash equilibrium, but a refinement of the concept nown as the "perfect" Nash equilibrium, originally introduced by elten (see [14]). This refinement, designed primarily for multi-stage ames, has the effect of preventing the chooser from formulating ontingent plans about what he would choose in various situations hat include implausible bluffs which are not called in equilibrium. ithout this refinement, the set of Nash equilibria would be much larger.[4]

Second, while I was preparing [3] for publication, I learned hat results parallelling some of mine had already been published y Kolm [11]. In [11], Kolm notes that the divide-and-choose game eed not always generate a Pareto-efficient allocation and, assuming convexity of preferences, shows that the outcome of the game is a fair allocation, that the role of divider is an advantage, and that the game treats agents equally and leads to an efficient allocation when preferences are identical. The parallel results presented in "A Game of Fair Division" improve upon these because they do not require convexity assumptions; in each case, establishing the more general results requires mathematical arguments that are essentially different, and that are, in my opinion, more illuminating than Kolm's. But he, of course, deserves full credit for priority.

The years since 1976 have made my original acknowledgements, if anything, more appropriate than when they were written. I am grateful to Dorothea Smits for additional typing and manuscript preparation. And in conclusion, I would like to add my thanks for the helpful subsequent comments of John Conlisk, Peter Hammond, Mark Machina, Andrew Postlewaite, Joel Sobel, anonymous referees, and participants in a seminar presentation at the University of California, San Diego.

La Jolla, California
July 20, 1980

FOOTNOTES

1 This essay extends results I obtained for two-person constant-sum games in [2].

2 A considerably shortened version of this essay appeared as [3].

3 Among the results in [9] is a partial characterization of when, in the presence of indivisibilities, the divide-and-choose method favors the chooser instead of the divider. Aware that chooser's advantage was a definite possibility when insufficient amounts of the divisible commodities are present, I suggested to my co-author that we use the method to allocate our reprints of [9]. He agreed, and was persuaded to be the divider, proceeding to spend several minutes measuring the two bundles against each other. I then chose, and received 31 reprints to his 22, a striking empirical confirmation of Theorem 2.1 in [9].

4 The chooser could, for example, formulate a contingent strategy that called for him to choose "irrationally" when faced with some of the divider's feasible divisions. If the divider's best response to this does not call into play the irrational part of the chooser's strategy, the strategies described are in imperfect Nash equilibrium. When the divider has convex preferences, for example, the set of

(perfect or imperfect) Nash equilibrium outcomes can be shown to equal the set of fair allocations that are also at least as good as equal division for the divider. The set of perfect Nash equilibrium outcomes is generally a unique point on the boundary of this set, the point most preferred by the divider.

REFERENCES

[1] Arrow, K.J.: "The Property Rights Doctrine and Demand Revelation Under Incomplete Information," Institute for Mathematical Studies in the Social Sciences, Stanford University, Technical Report 243, August 1977.

[2] Crawford, V.P.: "Learning the Optimal Strategy in a Zero-Sum Game," _Econometrica_, 42 (1974), 885-891.

[3] Crawford, V.P.: "A Game of Fair Division," _Review of Economic Studies_, 44 (1977), 235-247.

[4] Crawford, V.P.: "A Procedure for Generating Pareto-Efficient Egalitarian-Equivalent Allocations," _Econometrica_, 47 (1979), 49-60.

[5] Crawford, V.P.: "On Compulsory-Arbitration Schemes," _Journal of Political Economy_, 87 (1979), 131-159.

[6] Crawford, V.P.: "A Self-administered Solution of the Bargaining Problem," _Review of Economic Studies_, 47 (1980), 385-392.

[7] Crawford, V.P.: "Arbitration and Conflict Resolution in Labor-Management Bargaining," to appear in _American Economic Review Papers and Proceedings_, May 1981.

[8] Crawford, V.P.: "Maximin Behavior and Efficient Allocation," University of California, San Diego Discussion Paper 79-12.

[9] Crawford, V.P. and W.P. Heller: "Fair Division with Indivisible Commodities," Journal of Economic Theory, 21 (1979), 10-27.

[10] Donn, C.B.: "Games Final-Offer Arbitrators Might Play," Industrial Relations, 16 (1977), 306-314.

[11] Kolm, S.-C.: Justice et Équité, Paris: Éditions du Centre National de la Recherche Scientifique, 1972.

[12] Maskin, E.: "Implementation and Strong Nash Equilibrium," in Laffont, J. - J.(ed.), Aggregation and Revelation of Preferences, Amsterdam: North-Holland, 1979.

[13] Pazner, E., and D. Schmeidler: "Egalitarian-Equivalent Allocations: a New Concept of Economic Equity," Quarterly Journal of Economics, 92 (1978), 671-687.

[14] Selten, R.: "Reexamination of the Perfectness Concept for Equilibrium Points in Extensive Games," International Journal of Game Theory, 4 (1975), 25-55.

[15] Wilson, R.: "Competitive Exchange," Econometrica, 46 (1978), 577-585.

LEARNING BEHAVIOR AND THE NONCOOPERATIVE EQUILIBRIUM

1. INTRODUCTION

Nash's [13] noncooperative equilibrium, in which each agent optimizes given fixed conjectures about his opponents' actions, has long been the leading solution concept in economic models of interactive situations in which agents cannot make binding cooperative agreements. In [13], Nash demonstrated that, while a Nash equilibrium in pure (or unrandomized) strategies did not always exist in game situations, an equilibrium could be shown to exist under very weak assumptions if the concept of strategy was generalized to allow mixed strategies (or randomizations over the set of feasible pure strategies).

Since Nash wrote, his equilibrium concept has been carefully examined and extensively applied. We have recently been cogently reminded by Roberts and Sonnenschein [16] and Dasgupta and Maskin [6] that existence cannot be guaranteed in standard interactive economic models without allowing mixed strategies (or using other devices). These studies have led to a reexamination of the descriptive rationale for the Nash equilibrium, partly motivated, I suspect, by the hope that a convincing rationale would suggest an appropriate generalization of the concept for situations where (as in monopolistic competition) pure-strategy existence may fail but mixed strategies are thought to be inappropriate.

The traditional rationale stems from von Neumann and Morgenstern's [23] simplifying assumptions that game-theoretic agents are fully informed and completely rational, in effect possessing unlimited computational ability. For such agents, models that prescribe behavioral norms are also descriptive. And since there is some normative basis for using equilibrium strategies, especially in two-person constant-sum games, there is an equally strong basis for using the noncooperative equilibrium to describe behavior. In fact, a somewhat more sophisticated rationale is available along these lines, based on the currently popular "as-if" justifications for rational-expectations optimizing models. In this view, agents do not compute, but they do formulate expectations about what their opponents will do. To put the argument in its simplest form, assume that these are point expectations and that the game always has a unique Nash equilibrium. Then, agents' expectations will all be realized if and only if the actions they expect are in Nash equilibrium.[1]

In view of the relative weakness of the "rational-expectations" justification and the extent to which the study of noncooperative equilibria has dominated the literature applying game theory in economics, it is hardly surprising that game theorists began seeking alternative descriptive justifications for Nash's equilibrium concept, along the lines of those normally available for non-interactive optimizing models, very soon after the publication of <u>Theory of Games and Economic Behavior</u>. (See, for example, Arrow and Hurwicz [1,2], Rosen [18], Brown [3], Brown and von Neumann [4], Rapoport [15], Robinson [17], and von Neumann [22].) These studies typically assume that agents repeatedly play a static game, ignoring strategic interactions between different plays, and adjust their strategy choices according to some adaptive scheme, usually a gradient-type process in discrete or continuous time.

Under assumptions typically stronger than those used by Nash [13] to prove existence, these studies find some support for the belief that unsophisticated and imperfectly informed agents can "learn" their equilibrium strategies, at least in some game situations. But an interesting question they leave unresolved is whether the mixed-strategy equilibrium, which Nash showed to be a successful generalization of the pure-strategy equilibrium from the standpoint of proving existence, is also compatible with the adaptive justifications just described. If not, its use as a descriptive model requires an appeal to the rational-expectations justification,which seems considerably less plausible, especially in the light of the frequently counter-intuitive nature and well-known difficulty of computing equilibrium mixed strategies.

The question remains unresolved, in spite of the useful studies mentioned above, because Arrow and Hurwicz, and Rosen, employ an assumption of strict concavity of the payoff function, which, as they are aware, rules out mixed strategies. And while the earlier studies of Brown, Robinson, and von Neumann are specifically concerned with mixed strategies and obtain convergence to the mixed-strategy equilibrium in zero-sum two-person games, their primary goal was improving the computational techniques. As I pointed out in [5], their behavioral assumptions are descriptively less plausible as a result; [5] also criticizes Rapoport's study, which is closest in spirit to the present essay.

In spite of the above, it remains a plausible conjecture that convergence to mixed-strategy equilibria can be obtained in a behaviorally plausible adaptive model. As Dasgupta and Maskin [6, p. 12] put it: "A major criticism of the foregoing [iterative descriptive justification of the Nash equilibrium] is that the iterative process may not converge at all. Here, it is conceivable that mixed strategies may offer a definite advantage over pure ones, in that by convexifying the reaction correspondences, convergence might be facilitated. This last point, though, remains conjectural."[2] The purpose of this essay is to resolve this conjecture by extending a result proved for zero-sum two-person games in [5] to non-zero-sum n-person games.

Specifically, it is assumed that agents repeatedly play a static game in which they are uncertain about the consequences of their strategy choices, but learn about these consequences through experience. They use behaviorally plausible rules of thumb to adapt their choices of strategy in response to new information. The set of equilibria of the dynamic system that reflects these assumptions coincides with the set of (pure- or mixed-strategy) Nash equilibria of the static game they are playing. This system is, therefore, a natural candidate for an algorithm to compute equilibria. In fact, Arrow and Hurwicz [1,2] and Rosen [18] have shown that closely related algorithms converge to pure-strategy equilibria under reasonable, but not vacuous, assumptions. I shall show, however, that for "almost all" static games with mixed-strategy Nash equilibria, all members of a wide class of plausible adaptive mechanisms are locally unstable at equilibrium, and thus that Dasgupta and Maskin's [6, p.12] conjecture is false. An interesting analogy is provided to make clear why this is true. As it happens, there is a perfect formal analogy between the dynamic system described above and a discrete Walrasian tâtonnement. Mixed-strategy probabilities correspond to prices, and at a mixed-strategy equilibrium, the function that corresponds to the excess demand for a good is, except at the boundaries, independent of its "own" price; in "almost all" such economies, the tâtonnement process must be unstable. This is not the case at pure-strategy equilibria, because there, independence does not hold.

In Section 2 of this essay I shall present the model and prove the basic theorem, which applies directly only to games whose mixed-strategy solutions have all positive probabilities. In Section 3 I shall argue that the results of Section 2 also apply to "almost all" games whose mixed-strategy solutions include some zero probabilities as well. Finally, Section 4 will summarize the results of some simulation experiments with the model and Section 5 will discuss the implications of the result.

2. A SIMPLE MODEL OF LEARNING BEHAVIOR

Consider n players repeatedly playing a discrete noncooperative game whose solution is in mixed strategies with all positive probabilities. They have no direct knowledge of the payoff function, but are allowed to record their payoff together with the pure strategy actually played after each play of the game. They pause after each group of P plays to evaluate their returns and adjust their strategy mixtures, in an attempt to improve their returns if possible. P is taken to be sufficiently large for the law of large numbers to make the difference between average returns and their expectations negligible; this implies that information from still earlier periods is obsolete. It is highly unlikely that this simplification will affect the qualitative behavior of the models studied here.

Under these conditions, it is reasonable to assume that each player adjusts the probability with which he plays each pure strategy in response to the difference between the average return from that strategy and the average return from the entire game over the period just completed. Formally, at time t player i assigns the probabilities $z_t^i \equiv (z_{1t}^i, \ldots, z_{m_i-1t}^i)$ to the first m_i-1 of his m_i pure strategies. Let $z_t \equiv (z_t^1, \ldots, z_t^n)$, and define

$$\theta_j^i(z_t) \equiv \lambda_j^i(z_t) - H^i(z_t), \qquad (i=1,\ldots,n; \quad j=1,\ldots,m_i), \tag{2.1}$$

where $\lambda_j^i(z_t)$ is player i's expected gain from playing his j^{th} pure strategy, and $H^i(z_t)$ is his expected gain from an entire play of the game. Let the functions F^i be any sign-preserving functions that satisfy $F^i(0) = 0$ and are analytic throughout some neighborhood of the origin, and define

$$G^i[\theta_j^i(z_t), z_{jt}^i] \equiv \text{Max } \{F^i[\theta_j^i(z_t)], -z_{jt}^i\}, \tag{2.2}$$

$$(i=1,\ldots,n; \quad j=1,\ldots,m_i).$$

Then the system of nonlinear difference equations implied by the above assumptions can be written as follows:

$$(2.3) \qquad z^i_{jt} = \frac{z^i_{jt-1} + G^i[\theta^i_j(z_{t-1}),\ z^i_{jt-1}]}{1 + \sum_{k=1}^{m_i} G^i[\theta^i_k(z_{t-1}),\ z^i_{kt-1}]}, \quad (i=1,\ldots,n;\ j=1,\ldots,m_i)$$

(2.3) states that the mixed-strategy probability z^i_j is adjusted in response to θ^i_j, the difference between the average return from player i's j^{th} pure strategy and the average return from the entire game. Appropriate modifications are made to deal with boundaries, and the probabilities are deflated to insure that they sum to unity.[3] This system is closely related to the one used by Nash [13] in his fixed-point argument for the existence of a noncooperative equilibrium. Substituting from (2.2) shows that the numerator of the RHS of (2.3) is nonnegative. The denominator is strictly positive, since, by (2.2),

$$(2.4) \qquad 1 + \sum_{k=1}^{m_i} G^i[\theta^i_k(z_{t-1}),\ z^i_{kt-1}] \geq 1 - \sum_{k=1}^{m_i} z^i_{kt-1} = 0\ , \qquad (i=1,\ldots,n),$$

where the inequality is strict unless $F^i[\theta^i_j(z_{t-1})] \leq -z^i_{jt-1}$ for all $j=1,\ldots,m_i$. But multiplying (2.1) by z^i_{jt}, summing over j, and recalling the definition of $H^i(z_t)$ yields

$$(2.5) \qquad \sum_{j=1}^{m_i} z^i_{jt}\theta^i_j(z_t) \equiv 0, \qquad (i=1,\ldots,n).$$

Evaluating the sum in (2.5) at time t-1 and recalling that F^i is a sign-preserving function shows that it is not possible for $F^i[\theta^i_j(z_{t-1})] \leq -z^i_{jt-1}$ for all $j=1,\ldots,m_i$, so the inequality in (2.4) must be strict, and the denominator of the RHS of (2.3) is strictly positive. Thus, if the probabilities in z^i_{t-1} are nonnegative and sum to unity, the vector z^i_t must also lie in the unit simplex.

A sensible algorithm for computing noncooperative equilibria must stop at, and only at, noncooperative equilibria. The following lemma shows that the system (2.3) possesses this desirable property:[4]

Lemma 2.1: <u>The set of noncooperative equilibria coincides with the set of equilibria of (2.3).</u>

Proof: First suppose that $\hat{z}$, partitioned like z_t, is a noncooperative equilibrium. For any (i,j) such that $0 < \hat{z}^i_j < 1$, it must be the case that $\theta^i_j(\hat{z}) = 0$, since otherwise it is clear from (2.1) that player i could increase his expected payoff from the game by changing that component of his mixed strategy, a contradiction. By the same argument, for (i,j) such that $\hat{z}_j^i = 0$ or $\hat{z}_j^i = 1$, it must be the case that $\theta_j^i(\hat{z}) \leq 0$ or $\theta_j^i(\hat{z}) \geq 0$ respectively. If no component of $\hat{z}^i$ equals unity, it is clear from (2.2) that $G^i[\theta_j^i(\hat{z}), \hat{z}_j^i] = 0$ for all $j=1,\ldots,m_i$, so there is no change in player i's mixed strategy. If $\hat{z}_j^i = 1$ for some j, the fact that $G^i[\theta_k^i(\hat{z}), \hat{z}_k^i] = 0$ for all $k=1,\ldots,m_i$; $k \neq j$ implies that there is again no change in player i's mixed strategy. Since this argument can be repeated for each $i=1,\ldots,n$, the non-cooperative equilibrium $\hat{z}$ is indeed an equilibrium of the system (2.3).

To complete the proof of Lemma 2.1, suppose that $\hat{z}$ is an equilibrium of (2.3). (2.3) can then be rewritten as follows:

$$\text{(2.6)} \qquad \hat{z}_j^i \sum_{k=1}^{m_i} G^i[\theta_k^i(\hat{z}), \hat{z}_k^i] = G^i[\theta_j^i(\hat{z}), \hat{z}_j^i], \qquad (i=1,\ldots,n;\ j=1,\ldots,m_i).$$

If $\hat{z}_j^i = 0$, (2.6) implies that $G^i[\theta_j^i(\hat{z}), \hat{z}_j^i] = 0$ and thus that $\theta_j^i(\hat{z}) \leq 0$. Let S^i be the set of all indices j such that $0 < \hat{z}_j^i \leq 1$. (2.6) implies that for all j in S^i the sign of $G^i[\theta_j^i(\hat{z}), \hat{z}_j^i]$ is the same as that of the sum on the LHS of (2.6). But if $G^i[\theta_j^i(\hat{z}), \hat{z}_j^i] > 0$ for all j in S^i, $F^i[\theta_j^i(\hat{z})]$ and therefore $\theta_j^i(\hat{z})$ must be positive for all j in S^i, contradicting (2.5). Similarly, the assumption that $G^i[\theta_j^i(\hat{z}), z_j^i] < 0$ for all j in S^i leads to a contradiction, so it must be the case that $G^i[\theta_j^i(\hat{z}), \hat{z}_j^i] = 0$ for all j in S^i, and thus that $\theta_j^i(\hat{z}) = 0$. (While the Kuhn-Tucker condition requires only that $\theta_j^i(\hat{z}) \geq 0$ at a noncooperative equilibrium where $\hat{z}^i_j = 1$, $\theta^i_j(\hat{z}) = 0$ at such a point because of the definition of H^i.) I have argued that the equilibrium of (2.3) $\hat{z}$ satisfies the Kuhn-Tucker necessary conditions for the maximization problems that define a noncooperative equilibrium. Since, as is well known, the maximized functions $H^i(z_t)$ are linear in z^i_t, these necessary conditions are also sufficient. Thus $\hat{z}$ is an equilibrium of (2.3) if and only if it is a noncooperative equilibrium. $\square$

We are now ready to prove the main result. Let us ignore boundary problems for the present by assuming that all components of the noncooperative equilibrium $\hat{z}$ are strictly positive; this assumption will be relaxed in Section 3. In this case, throughout a neighborhood of $\hat{z}$,

$$(2.7) \qquad G^i[\theta^i_j(z_t), z^i_{jt}] \equiv F^i[\theta^i_j(z_t)], \quad (i=1,\ldots,n;\ j=1,\ldots,m_i).$$

Partially differentiating the identity (2.1) with respect to z^i_{ht} and evaluating the result at $\hat{z}$ yields:

$$(2.8) \qquad \frac{\partial\theta^i_j(\hat{z})}{\partial z^i_{ht}} = 0, \quad (i=1,\ldots,n;\ j=1,\ldots,m_i),$$

because $\lambda^i_j(z_t)$ is independent of z^i_t by definition, and $H^i(z_t)$

reaches an interior maximum with respect to z_t^i at $\hat{z}$.[5] Since all components of $\hat{z}$ are strictly positive by assumption, the proof of Lemma 2.1 shows that

$$\theta_j^i(\hat{z}) = 0, \quad (i=1,\ldots,n; \quad j=1,\ldots,m_i). \tag{2.9}$$

Partial differentiation of (2.3) and substitution from (2.8) and (2.9) yields

$$\frac{\partial z_{jt}^i}{\partial z_{jt-1}^i} = 1, \quad (i=1,\ldots,n; \quad j=1,\ldots,m_i-1), \tag{2.10}$$

and

$$\frac{\partial z_{jt}^i}{\partial z_{ht-1}^i} = 0, \quad (i=1,\ldots,n; \quad h,j=1,\ldots,m_i-1; \quad h\neq j), \tag{2.11}$$

where the partial derivatives in (2.10) and (2.11) are evaluated at the equilibrium $\hat{z}$. Thus, the matrix, A, of the linear system associated with (2.3) at its equilibrium has identity matrices along its main diagonal, so that

$$\text{tr } A = \sum_{i=1}^{n} (m_i-1), \tag{2.12}$$

the number of independent

variables in the system. Since tr A also equals the sum of the characteristic roots of A, this implies that either all the characteristic roots of A are equal to unity or that at least one of them has a modulus greater than unity. In the latter case the associated linear system is unstable, and thus the system (2.3) is locally unstable. In the former case, if the characteristic vectors of A are not independent the solution of the associated linear system has a polynomial time trend and is again unstable. If the characteristic vectors are independent and all the characteristic roots equal unity, then A = I. Even when all the characteristic roots equal unity, (2.3) is locally unstable except when all of the off-diagonal elements of A equal zero, which happens only on a set of measure zero.[6] This phenomenon occurs for zero-sum two-person games in, and only in, the trivial game where payoffs do not depend on players' actions. In more general games, A = I can hold for nontrivial games as well.

We have now established the following theorem:

Theorem 2:1: For all but a set of measure zero discrete noncooperative n-person games with solutions in strictly positive mixed strategies, the learning process described by the system (2.3) is locally unstable.

Even if I had followed Rapoport [15] in modeling the

learning process as a system of differential equations instead of difference equations, the conclusion of Theorem 2.1 would have been the same.

The system (2.3) is formally analogous to a Walrasian tâtonnement system where prices are normalized after each adjustment to lie on the unit simplex. The z^i_{jt} correspond to prices, and the $\theta^i_j(z_t)$, which are clearly continuous and differentiable, correspond to excess demand functions. Summing equation (2.5) over i yields a result that corresponds to Walras' Law, since (2.5) is an identity in z_t. It is well known that any system of community excess demand functions that satisfy Walras' Law and are continuous can be obtained by adding the demands of a finite number of appropriately chosen utility-maximizing, price-taking agents; see Sonnenschein [20] for example. Thus, the system (2.3) associated with a given game could also have arisen from a discrete Walrasian tâtonnement in an appropriately chosen competitive exchange economy. The instability of the present model arises because the functions that correspond here to excess demand functions are, except at the boundaries, independent of the mixed-strategy probabilities that correspond to their "own" prices.

As the results of Arrow and Hurwicz [1,2] and Rosen [18] and the literature on the stability of the Cournot oligopoly model (see Fisher [7] and the references therein) reveal, this instability need not extend to games with pure-strategy equilibria. The difference arises because the independence property just mentioned is the discrete analog of a zero second derivative of the payoff function. The second-order condition takes this form because of the linearity imposed by the von Neumann-Morgenstern "Independence" Axiom in the mixed-strategy case, which does not extend to the pure-strategy case in general.

3. EXTENSION TO THE GENERAL CASE

Theorem 2.1 would be of some interest even if nothing could be said about the more general case where some of the components of the equilibrium mixed strategies are zero. It is possible to argue, however, that the results of the previous section extend to all but a set of measure zero of the games in this class as well. In the discussion that follows, degenerate games having pure strategies that are assigned zero probabilities in the solution but have expected returns at the solution equal to the value of the game are excluded from consideration; this occurs only on a set of measure zero. The proof of Lemma 2.1 and the continuity of the λ_j^i show that in the games that remain, if a pure strategy is assigned a zero probability in the solution, that pure strategy must have below-average returns throughout a neighborhood of the solution.

It follows that players whose collective behavior can be described by (2.3) will never raise these probabilities above zero in the neighborhood of the solution, except possibly in the degenerate cases excluded from consideration above. The behavior of the system (2.3) is independent of the payoffs corresponding to zero-probability pure strategies, since these payoffs enter the functions θ_j^i only with zero weights. Therefore, for our purposes, we can ignore zero-probability pure strategies without loss of generality.[7] Theorem 2.1 can then be extended to the general mixed-strategy case as well:

Theorem 3.1: For all but a set of measure zero discrete noncooperative n-person games with solutions in mixed strategies, the learning process described by the system (2.3) is locally unstable.

4. SIMULATION EXPERIMENTS

In the special case of a zero-sum, two-person game with a 2x2 payoff matrix, I have simulated the model to learn about its global properties. In the simulations the stochastic nature of the system was taken into account. The results of a few simple experiments, originally reported in [5], can be summarized here:

(i) Pure-strategy saddle points are globally stable, at least in these simple games.[8]

(ii) After the initial departure from mixed-strategy equilibrium (induced by the failure of random payoffs to realize their expectations exactly), the model oscillates generally around the equilibrium, with amplitude increasing over time. The violence of these oscillations increases with the slopes of F^1 and F^2 as might be expected (these slopes are constant since F^1 and F^2 were linear in the experiments), and for moderate values of these slopes the zero/one probability boundaries become effective quite often.

(iii) In these cases, the boundaries cause a consistent distortion of the average observed mixed strategies toward the barycenters, $z_t^1 = (1/m_1,\dots,1/m_1)$ and $z_t^2 = (1/m_2,\dots,1/m_2)$, and in general the property of Rapoport's model that the observed probabilities average over time to the optimal strategies is lost.

5. CONCLUSIONS

In this essay I have investigated the possibility of players locating the mixed-strategy equilibrium of a noncooperative n-person game through an iterative learning process. All members of the class of such processes that I consider the most behaviorally plausible were shown to be locally unstable for "almost all" n-person games with mixed-strategy noncooperative equilibria. This instability corresponds to that which would be exhibited by a discrete Walrasian tâtonnement process if the excess demand for each good were independent of its own price at equilibrium. This instability need not extend to pure-strategy noncooperative equilibria.

Thus, while the mixed strategy is a successful generalization of the concept of strategy from the normative standpoint and for the purpose of insuring the existence of a noncooperative equilibrium, the arguments available to justify its use to describe behavior are considerably weaker than for pure strategies. Unless new learning models that are behaviorally plausible and have better convergence properties are developed, models with mixed-strategy Nash equilibria must rely, for their descriptive plausibility, on heroic assumptions akin to those used to justify the study of rational-expectations models.

FOOTNOTES

[1] To the best of my knowledge, this view has never been clearly stated in the literature, although it is implicit in many discussions of the Nash equilibrium. Harsanyi [9, pp. 163-165] seems to come closest to stating it explicitly, in the context of games with incomplete information. Townsend [21] and Prescott and Townsend [14] obtain similar results asymptotically (that is, as the number of agents grows) in models that differ essentially from those considered here in that they include market variables (prices, for example) that transmit information about actions across agents.

Harsanyi and Selten's [10, pp. P-89-P-92] discussion suggests how to generalize the "if and only if" statement in the text to situations where the equilibrium is not unique.

[2] Green and Stokey [8, pp. 10-11], however, conjecture the opposite result.

[3] Allowing one of each player's probabilities to be determined as a residual is simpler but asymmetrical, and leads to the same ultimate conclusions.

[4] Of course, Nash [13] had to prove an analogous result, but the difference between his system and (2.3) makes a new argument necessary.

[5] Note that the arguments of H^i include only the first m_i-1 components of player i's mixed strategy vector; the constraint that probabilities must sum to unity has already been substituted out and the nonnegativity constraints are not binding, so $\partial H^i(\hat{z})/\partial z^i_{ht} = 0$ is the appropriate first-order condition.

[6] For the argument that A=I in this case, see McManus [12]. To see that the off-diagonal elements of A are all zero only on a set of measure zero, note first that when an equilibrium has all positive mixed-strategy probabilities, any player can unilaterally change his strategy to any other strategy without destroying equilibrium. Thus, games with A=I necessarily have an infinite number of equilibria; but the set of such games is of measure zero.

7 There is a subtle problem of interpretation here, since it has been assumed that players have information about the average returns of pure strategies even when they are played only with zero probability in equilibrium. Formally, this is clearly wrong, but not seriously so for my purposes. Local stability analysis examines the effect of perturbing a system a small distance away from equilibrium, but its interest lies in the information it yields about whether a system, approaching an equilibrium, will be attracted or repelled. Players may be able to collect information about average returns along a path approaching equilibrium, even though the information is no longer available once equilibrium is reached.

8 In fact, it is possible to prove this when the payoff matrix is 2x2.

REFERENCES

[1] Arrow, K. J., and L. Hurwicz: "Gradient Method for Concave Programming, I: Local Results and III : Further Global Results and Application to Resource Allocation," Chapters 6 and 8 in Arrow, K. J., L. Hurwicz, and H. Uzawa (eds.), Studies in Linear and Nonlinear Programming, Stanford California: Stanford University Press, 1958.

[2] Arrow, K. J., and L. Hurwicz: "Stability of the Gradient Process in N-Person Games," Journal of the Society for Industrial and Applied Mathematics, 8 (1960), 280-294; reprinted in Arrow, K. J., and L. Hurwicz (eds.), Studies in Resource Allocation Processes, London and New York: Cambridge University Press, 1977.

[3] Brown, G. W.: "Iterative Solutions of Games by Fictitious Play," in Koopmans, T. C. (ed.), Activity Analysis of Production and Allocation, Cowles Commission Monograph 13, New York: John Wiley and Sons, Inc., 1951.

[4] Brown, G.W., and J. von Neumann: "Solution of Games by Differential Equations," in Kuhn, H. W., and A. W. Tucker (eds.), Contributions to the Theory of Games, Vol. 1, Annals of Mathematics Studies #24, Princeton, New Jersey: Princeton University Press, 1950.

[5] Crawford, V.P.: "Learning the Optimal Strategy in a Zero-Sum Game," Econometrica, 42 (1974), 885-891.

[6] Dasgupta, P., and E. Maskin: "The Existence of Economic Equilibria: Continuity and Mixed Strategies," Technical Report 252, Institute for Mathematical Studies in the Social Sciences, Stanford University, November 1977.

[7] Fisher, F. M.: "The Stability of the Cournot Oligopoly Solution: the Effects of Speeds of Adjustment and Increasing Marginal Costs," Review of Economic Studies, 28 (1961), 125-135.

[8] Green, J. R., and N. L. Stokey: "A Two-Person Game of Information Transmission," Discussion Paper 418, Northwestern University, March 1980.

[9] Harsanyi, J. C.: "Games with Incomplete Information Played by 'Bayesian' Players I: The Basic Model," Management Science, 14 (1967), 159-182.

[10] Harsanyi, J.C., and R. Selten: "A Generalized Nash Solution for Two-Person Bargaining Games with Incomplete Information," Management Science, 18 (1972), P80-P106.

[11] Luce, R. D., and H. Raiffa: Games and Decisions: Introduction and Critical Survey, New York: John Wiley and Sons, Inc., 1957.

[12] McManus, M.: "Dynamic Cournot-Type Oligopoly Models - a Correction," Review of Economic Studies, 29 (1962), 337-339.

[13] Nash, J.F.: "Non-Cooperative Games," Annals of Mathematics, 54 (1951), 286-295.

[14] Prescott, E., and R. Townsend: "Equilibrium Under Uncertainty: Multi-Agent Statistical Decision Theory," to appear in Zellner, A. (ed.), Studies in Bayesian Econometrics and Statistics in Honor of Harold Jeffreys.

[15] Rapoport, A.: Two Person Game Theory: the Essential Ideas, Ann Arbor, Michigan: The University of Michigan Press, 1966.

[16] Roberts, J., and H. Sonnenschein: "On the Foundations of the Theory of Monopolistic Competition," Econometrica, 45 (1977), 101-113.

[17] Robinson, J.: "An Iterative Method of Solving a Game," Annals of Mathematics, 54 (1951), 296-301.

[18] Rosen, J.: "Existence and Uniqueness of Equilibrium Points for Concave N-Person Games," Econometrica, 33 (1965), 520-534.

[19] Samuelson, P. A.: Foundations of Economic Analysis, New York: Atheneum, 1970.

[20] Sonnenschein, H.: "Do Walras' Law and Continuity Characterize the Class of Community Excess Demand Functions?", Journal of Economic Theory, 6 (1973), 345-354.

[21] Townsend, R.: "Market Anticipations, Rational Expectations, and Bayesian Analysis," International Economic Review, 19 (1978), 481-494.

[22] von Neumann, J.: "A Numerical Method to Determine Optimum Strategy," Naval Research Logistics Quarterly, 1 (1954), 109-115.

[23] von Neumann, J., and O. Morgenstern : Theory of Games and Economic Behavior, New York: John Wiley and Sons, Inc., 1947.

A GAME OF FAIR DIVISION

A chap out of the Illinois River, with a little stern-wheel tub, accosts a couple of ornate and gilded Missouri River pilots:

"Gentlemen, I've got a pretty good trip for the up-country, and shall want you about a month. How much will it be?"

"Eighteen hundred dollars apiece."

"Heavens and earth! You take my boat, let me have your wages, and I'll divide!"

--Mark Twain, Life on the Mississippi

1. INTRODUCTION

The "divide and choose" method has played an important role in the literature on fair division.[1] This technique for allocating bundles of goods seems impartial, requires little cooperation from agents, and is nearly free of administrative costs. It is therefore somewhat puzzling that it has found so few applications in the real world, where sometimes even prolonged and costly negotiations produce only imperfect agreements. Either the method has drawbacks not yet well understood, or it is underutilized. This essay examines the game that arises when two agents agree to use the divide and choose method. The analysis leads to a resolution of the puzzle mentioned above, and identifies a class of situations where replacing conventional arbitration procedures with the divide and choose method can be strongly recommended.

In the sequel, an agent who would prefer another agent's bundle of goods to his own will be said to envy the other agent. An allocation at which no agent envies another will be called a fair allocation.[2] A well-known property of the two-person version of the divide and choose game[3] is that each player can insure that he does not envy the other. The divider (D) can accomplish this by dividing so that he is indifferent to his opponent's choice; the chooser (C) need only choose his most preferred bundle after D divides. That the players, acting together, can insure that the outcome of the game is fair is interesting, but more information about the allocations actually generated by the game is needed to judge its usefulness as a fair division device. Conceivably, with players motivated by self-interest, the game could generate an unfair allocation in spite of the above result.

To learn more about the divide and choose method, I assume that players seek to obtain the most desirable bundle possible. They are also assumed to behave noncooperatively, since negotiating a mutually acceptable settlement would be relatively easy if they were willing to cooperate, and the method would then be superfluous. In Section 2 of this essay D's problem is formulated and his optimal noncooperative strategy is characterized. As is suggested by Luce and Raiffa [6, p. 365] and Singer [9], the common belief that D should divide the bundle so that he is indifferent about C's choice is false. If D knows C's preferences with certainty, under very general conditions -- roughly, that players' behavior can be described by the maximization of continuous and strongly monotonic utility functions and that goods are homogeneous and perfectly divisible -- his optimal strategy involves dividing the bundle so that C is indifferent about his choice.

Once D's optimal strategy has been characterized, several interesting conclusions follow. Luce and Raiffa's belief [6, pp. 364-365] that the allocations resulting from this game are always Pareto-efficient is false. However, the noncooperative equilibrium of the game is a fair allocation and, under an additional mild behavioral assumption, is an efficient point in the set of all fair allocations. Luce and Raiffa's statement [6, p. 365] that the role of divider is an advantage in the game if preferences are known is formalized and proved, and the allocation resulting from the game is shown to treat agents less equally than an equal-income competitive equilibrium (EICE) allocation.[4]

When both agents have identical preferences, however, the Pareto-inefficiency, divider's advantage, and unequal treatment all disappear, and these results are robust in the sense that if agents have nearly identical preferences, nearly equal treatment and near-Pareto-efficiency result. Section 2 is self-contained, and the reader who has more faith than time may wish to read only Section 2 and Section 5, which includes a summary of the generalizations of Sections 3 and 4.

In Section 3, all but one of the pure-trade results (the one that establishes the relationship between game allocations and EICE allocations) are generalized to the case where production is possible -- that is, where each agent is endowed with productive resources that may either be consumed directly or used to produce other consumption goods. In this context, D plans the operation of the economy -- required factor supplies, production levels, and consumption shares -- and C then chooses which of the two roles in the economy he will assume.

In Section 4 the analysis is carried out in the case where the divide and choose process is followed by cooperative trade that both players anticipate in their strategy choices. Some of the results, notably the existence of divider's advantage, go through in this case. Finally, Section 5 discusses several possible further generalizations that would be of interest,

summarizes the results obtained, and considers their implications.

2. THE PURE-TRADE DIVIDE AND CHOOSE GAME

Assume that two persons have agreed to share a fixed bundle of homogeneous and perfectly divisible goods by the divide and choose method, and that their roles have already been determined in some way, perhaps by the toss of an unbiased coin. Each person behaves noncooperatively, seeks only to obtain the most desirable bundle possible, and has preferences that are representable by a continuous and strongly monotonic[5] (though not necessarily quasiconcave) utility function. In addition, D is assumed to know C's preferences with certainty, which is not a completely natural assumption to make in most situations. But studying the certainty case permits clearing up an area of confusion in the literature, provides a necessary preliminary to a more general analysis, and may serve as a reasonable description of some situations. Finally, an innocuous assumption is included to simplify the exposition. Whenever C is indifferent between the two bundles offered him he can be counted on to choose the one that D would prefer him to. Since it has already been assumed that C's preferences are known with certainty, this assumption is innocuous -- a tiny adjustment in D's division could induce C to make the desired choice without perceptibly altering either player's consumption or welfare. This assumption allows us to deal with maxima instead of suprema, and greatly simplifies the statements of some of the results.

In the sequel, the following vector notation is used: if $a \equiv (a_1,\ldots,a_r)$ and $b \equiv (b_1,\ldots,b_r)$, $a \leqq b$ means $a_i \leq b_i$, $(i=1,\ldots,r)$, $a \cdot b \equiv \sum_{i=1}^{r} a_i b_i$, and $ab \equiv (a_1 b_1,\ldots,a_r b_r)$. $\underline{0}$ and $\underline{1}$ denote a vector of zeros and a vector of ones, whose dimensionalities should be inferred from the context.

Units are chosen so that the vector of goods to be allocated is $\underline{1}$. A division by D will be represented by the m-vector z, indicating that C may choose between the consumption bundle z and its image $\underline{1}$-z. Let U^D and U^C be D's and C's utility functions, and assume without loss of generality that z is the bundle D intends for himself. Then D's optimal noncooperative strategy in the game is any solution of the following programming problem:

$$\text{(2.A)} \qquad \max_{\underline{0} \leqq z \leqq \underline{1}} U^D(z) \qquad \text{s.t. } U^C(z) \leq U^C(\underline{1}-z).$$

A solution to (2.A) always exists, since a continuous function defined on a compact, nonempty ($z = \frac{1}{2}\underline{1}$ is always feasible) set always takes on a maximum value at some point in the set. The following theorem provides an optimality condition that any solution must satisfy:

Theorem 2.1: Any solution to (2.A) must satisfy

(2.1) $U^C(z) = U^C(\underline{1}-z)$.

Proof: The proof is by contradiction. Suppose that $\tilde{z}$ is a solution to (2.A) and that $U^C(\tilde{z}) < U^C(\underline{1}-\tilde{z})$. If $\tilde{z} = \underline{1}$ we would have the contradiction $U^C(\underline{1}) < U^C(\underline{0})$, so at least one component of $\tilde{z}$ must be less than unity; renumber the goods if necessary so the first component is less than unity. Let $\tilde{z}[\varepsilon] \equiv \tilde{z} + (\varepsilon, 0,\ldots,0)$. By the continuity of U^C there exists an $\varepsilon > 0$ such that $\underline{0} \leqq \tilde{z}[\varepsilon] \leqq \underline{1}$, $U^C(\tilde{z}[\varepsilon]) \leq U^C(\underline{1} - \tilde{z}[\varepsilon])$, and $U^D(\tilde{z}[\varepsilon]) > U^D(\tilde{z})$. This contradicts the assumption that $\tilde{z}$ is a solution to (2.A), and establishes Theorem 2.1. □

The solution of (2.A) in the two-good case is illustrated in the Edgeworth box diagram of Figure 2.1. In the diagram are shown three of C's indifference curves (C_1, C_2, and C_3), one of D's (D_1), and the locus of points (EE) that satisfy (2.1). The EE locus can be constructed from C's indifference map as follows. First, it is clear that it passes through the point $\frac{1}{2}\underline{1}$ (A in Figure 2.1). Additional points on the locus can be generated by drawing a circle with center at A and finding a pair of antipodal points on the circle that lie on the same indifference curve. If

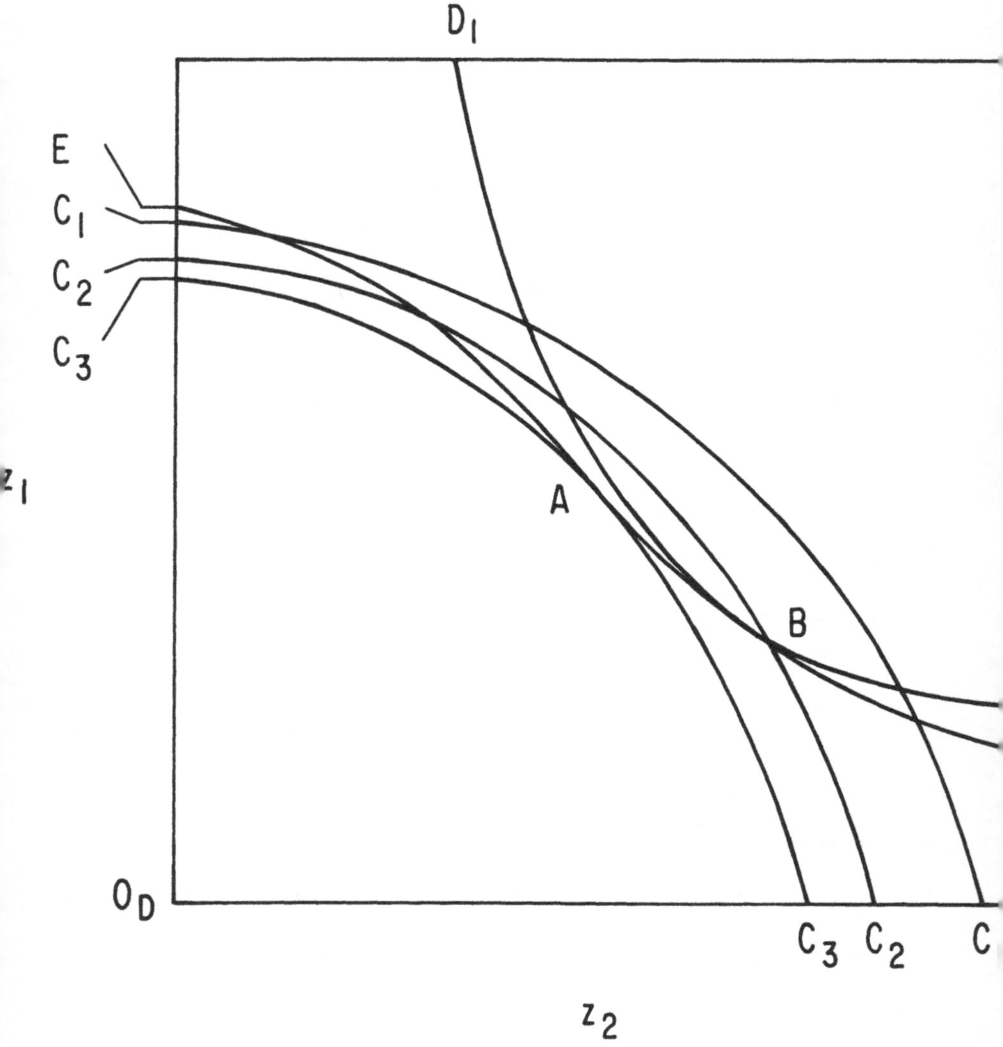

Figure 2.1

U^C is differentiable, total differentiation of (2.1) reveals that the EE locus has the same slope at A as C's indifference curve through that point. If, in addition, U^C is quasiconcave,

$$(2.2) \qquad U^C(\tfrac{1}{2}\underline{1}) \equiv U^C(\tfrac{1}{2}z'' + \tfrac{1}{2}[\underline{1} - z'']) \geq \tfrac{1}{2}U^C(z'') + \tfrac{1}{2}U^C(\underline{1} - z'') = U^C(z'') = U^C(\underline{1} - z''),$$

where z" is any point on the EE locus. Thus, if U^C is quasiconcave and differentiable, the EE locus lies entirely to the northeast of C_3, C's indifference curve through A, and must therefore be tangent to it at that point.

All points in the box that are not to the northeast of the EE locus satisfy the constraints of (2.A). If D makes a division represented by one of these points, C will choose so that the allocation represented by that point results. Naturally, D chooses the feasible point that yields him the greatest utility (B in Figure 2.1). (If D made a division represented by a point to the northeast of the EE locus, C would choose so that the allocation represented by the image of that point resulted. The outcome would therefore be the same as if D had made the division represented by the image point in the first place.)

It is evident from Figure 2.1 that the allocation of goods resulting from the game need not be Pareto-efficient in general. The following example shows this directly, and may help to illustrate the way the game works. Suppose that there are only two goods, z_1 and z_2, and let $U^D(z_1,z_2) \equiv (z_1+1)(z_2+2)$ and $U^C(z_1,z_2) \equiv z_1 + z_2 + (z_1z_2)^{\frac{1}{2}}$. Both of these utility functions are strictly quasiconcave and strictly increasing in z_1 and z_2, so there is nothing special about this example. The reader may verify (noting that, by the symmetry and monotonicity of U^C in z_1 and z_2, the optimality condition (2.1) is equivalent to the condition $z_1 + z_2 = 1$ in this case) that D's optimal division is (1,0). D obtains this bundle and a utility of 4 in the game, while C obtains the bundle (0,1) and a utility of 1. If C then gave D 2ε units of z_2 in exchange for ε units of z_1, D's utility would increase to $4+2\varepsilon-2\varepsilon^2 > 4$, and C's would increase to $1-\varepsilon+(\varepsilon-2\varepsilon^2)^{\frac{1}{2}} > 1$, where both inequalities hold whenever $0 < \varepsilon < \frac{1}{3}$. Since both players would gain from such a trade, the original allocation could not have been Pareto-efficient.

Before continuing with the main results of this section, it is convenient to pause for a lemma:

Lemma 2.1: <u>By dividing appropriately, D can enforce any feasible fair allocation as the outcome of the pure-trade divide and choose game.</u>

Proof: An allocation giving z to D and $\underline{1}-z$ to C is fair if and only if $U^D(z) \geq U^D(\underline{1}-z)$ and $U^C(z) \leq U^C(\underline{1}-z)$.

By inspection, the region defined by these two constraints and the feasibility condition $0 \leq z \leq 1$ is entirely contained in the feasible region of (2.A). □

The next theorem establishes that the allocations generated by the game are fair and that, although they need not be Pareto-efficient, they are efficient in a weaker sense. There may exist allocations that both players would prefer to the outcome of the game, but there are no such fair allocations.

Theorem 2.2: The outcome of the pure-trade divide and choose game is a fair allocation, and if D agrees to break ties for the solution of (2.A) by dividing as C would prefer, the outcome is an efficient point in the set of all fair allocations.

Proof: The first part of the theorem follows directly from Theorem 2.1 and our assumption about C's choice when he is indifferent.[6] To prove the second part, observe that, by Lemma 2.1, D can enforce any fair allocation, so there cannot be any fair allocations that D prefers to the outcome of the game. Thus, either the outcome is efficient in the set of fair allocations or there exists a fair allocation that yields D the same utility as the bundle he obtains in the game, and that C prefers to the outcome of the game. But the latter possibility is ruled out by our assumption about D's tie-breaking, so the outcome is efficient in the set of fair allocations. □

The next theorem formalizes the notion that the role of divider is an advantage in the game if preferences are known.

Theorem 2.3: <u>In the pure-trade divide and choose game, D always does at least as well in the role of divider as he would in the role of chooser.</u>

Proof:[7] Consider the game in which C divides and D chooses. The allocation generated by this game is fair by Theorem 2.2. Therefore, by Lemma 2.1, D can enforce this allocation when he is divider, and so must do at least as well in the role of divider as he would in the role of chooser. □

Luce and Raiffa's example [6, pp. 364-365] of the game when the bundle to be divided consists of a single indivisible object shows that some restrictions on goods or preferences are necessary for Theorem 2.3 to be true. Strictly speaking, the above proof shows that the role of divider can never be a disadvantage, but begs the question of whether it is ever, or always, an advantage. The argument used to establish (2.2) and the fact that $\frac{1}{2}\underline{1}$ is feasible for (2.A) show that the role of divider is an advantage when U^D is strictly quasiconcave unless $\frac{1}{2}\underline{1}$ happens to be both D's and C's optimal division. When U^D

is not quasiconcave, it is easy to construct examples to show that it can still be an advantage.

The next theorem compares the allocations generated by the divide and choose game with those from an EICE. An EICE is shown to treat agents more equally, in a sense made precise by the theorem, than the game does. Assume that an EICE exists in spite of possible nonquasiconcavities in U^D and U^C.

Theorem 2.4:[8] <u>In the pure-trade divide and choose game, D always does at least as well as he would at an EICE and C always does at most as well as he would at an EICE.</u>

Proof: A pure-trade EICE is necessarily fair -- no agent can envy another agent, because he could have afforded the other's bundle but chose not to buy it.[9] Therefore, by Lemma 2.1, D can always enforce an EICE allocation in the game, which establishes the first part of the theorem. To prove the second part of the theorem, note that, since all goods are desirable and we have just shown that D likes his game allocation at least as well as his EICE allocation, and the EICE allocation is Pareto-efficient, C must like his game allocation at most as well as his EICE allocation, which completes the proof. □

It is easy to verify that if U^C is quasiconcave equal division (ED) of the bundle also possesses the properties attributed to the EICE in Theorem 2.4.

The next two theorems show that, when D and C have identical preferences, the allocation generated by the divide and choose game treats them equally and is fully Pareto-efficient. In this case at least it is possible, even without cooperation, to achieve an allocation that is very good when judged by the traditional standards of equity and efficiency without incurring large administrative costs.

Theorem 2.5: If D and C have identical preferences, they are treated equally in the pure-trade divide and choose game.

Proof: D's and C's identical preferences can both be represented by the same utility function. By Theorem 2.1, both players derive the same utility from the game no matter who is the divider. □

Theorem 2.6: When D and C have identical preferences, the outcome of the pure-trade divide and choose game is a Pareto-efficient allocation.

Proof: The proof is by contradiction. Let ϕ represent both U^D and U^C, and suppose that $\bar{z}$ is a solution of problem (2.A) that is not Pareto-efficient. Then there exists a z' such that

$\underline{0} \leqq z' \leqq \underline{1}$,

(2.3) $\phi(z') \geq \phi(\bar{z})$, and

(2.4) $\phi(\underline{1}-z') \geq \phi(\underline{1}-\bar{z})$,

with strict inequality holding in at least one of (2.3) and (2.4). Suppose first that (2.3) holds with strict inequality. Then

(2.5) $\phi(z') > \phi(\underline{1}-z') \geq \phi(\underline{1}-\bar{z}) \geq \phi(\bar{z})$.

The first inequality in (2.5) follows from the fact that, because (2.3) holds with strict inequality, z' must not be feasible for (2.A); the second from (2.4); and the third from the fact that $\bar{z}$ is feasible for (2.A).

It follows from (2.5) that $\underline{1} - z'$ is feasible for (2.A). If $\phi(\underline{1}-z') > \phi(\bar{z})$, this yields a contradiction immediately. If not, $\phi(\underline{1}-z') = \phi(\bar{z})$ and $z' \neq \underline{0}$ by (2.5); renumber the goods if necessary so the first component of z' is strictly positive. Let $z'[\varepsilon] \equiv z' - (\varepsilon,0,\ldots,0)$. By the continuity and monotonicity of ϕ there exists an $\varepsilon > 0$ such that $\underline{0} \leqq z'[\varepsilon] \leqq \underline{1}$, $\phi(\underline{1}-z'[\varepsilon]) \leq \phi(z'[\varepsilon])$, and $\phi(\underline{1}-z'[\varepsilon]) > \phi(\bar{z})$, contradicting the hypothesis that $\bar{z}$ was a solution of (2.A).

To complete the proof, suppose instead that it is (2.4) that holds with strict inequality. Then

(2.6) $\phi(\underline{1}-z') > \phi(\underline{1}-\bar{z}) \geq \phi(\bar{z})$.

If $\phi(z') \geq \phi(\underline{1}-z')$, $\underline{1} - z'$ is feasible for (2.A) and strictly preferred to $\bar{z}$ by (2.6), a contradiction. If $\phi(z') < \phi(\underline{1}-z')$, z' is feasible for (2.A), a contradiction unless (2.3) holds with equality. Finally, if (2.3) does hold with equality, a perturbation of z' yields a vector that is both feasible for (2.A) and strictly preferred to $\bar{z}$; this final contradiction completes the proof. □

Theorem 2.6 is immediate when preferences are convex because the outcome of the game then coincides with ED, which is Pareto-efficient when preferences are convex and identical, and the EICE, which is always Pareto-efficient. However, Theorem 2.6 is a more general result: when preferences are identical the game always generates a Pareto-efficient allocation, even, for example, when the EICE fails to exist due to nonconvexities.

Theorems 2.5 and 2.6 attribute two highly desirable properties to the divide and choose method when D and C have identical preferences: equal treatment and Pareto-efficiency. These properties hold necessarily whenever there is a single, desirable good in the bundle to be allocated. In this case, agents' preferences may be represented by identical, quasi-concave utility functions for my purposes, which require only the ordinal properties of preferences, and Theorems 2.5 and 2.6 therefore apply directly. However, it is extremely unlikely that the assumption of identical preferences should be literally realized in the general, m-good case -- even identical twins' tastes diverge as they are formed by different experiences. It would therefore be of interest to know if the results of Theorems 2.5 and 2.6 are robust in the sense that, when agents have nearly identical preferences, the game treats them nearly equally and generates a nearly Pareto-efficient allocation.

Fortunately, this is the case, although the results are not trivial. Even if preferences are convex the feasible region of (2.A) is not generally convex, so there is no economically attractive way to rule out discontinuous changes in the allocation resulting from the game in response to changes in preferences. However, the next two theorems establish that, at least when preferences are identical, these discontinuous changes do not induce discontinuities in the utility outcomes of the game. It follows that players with nearly identical preferences are treated nearly equally in the game, and since a small change in preferences induces a small change in the utility possibility frontier, when players have nearly identical preferences the game generates a nearly Pareto-efficient allocation.

It will be convenient to introduce a slightly different notation. Assume that both agents have preferences that are representable by a member of the same p-parameter family of utility functions, where p may be large, but must be finite.[10] Let I^q denote the q-fold Cartesian product of the closed unit interval I with itself, and let U(d,z) and U(c,z) denote D's and C's utility functions, where $z \in I^m$ is a consumption vector as before, and $d \in I^p$ and $c \in I^p$ are vectors of parameters. Further assume that U is jointly continuous in its arguments, define

$$(2.7) \qquad V^D(c,d) \equiv \underset{z \in I^m}{\text{Max}} \; U(d,z) \qquad \text{s.t.} \quad U(c,z) \leq U(c,\underline{1}-z),$$

and call the problem in (2.5) "(2.B)". Since U is continuous and the constraints of (2.B) define a closed nonempty set, Berge's Maximum Theorem[11] implies the following result:

Theorem 2.7: <u>V^D is everywhere jointly continuous in c and d.</u>

Theorem 2.7 is intuitively plausible. Wherever D's optimal division varies continuously with preferences, V^D is also continuous. As we have seen above, discontinuities in the optimal division cannot be ruled out, but even if the optimal policy jumps from one point to another D must be indifferent between these points, so such jumps do not induce discontinuities in V^D.

The continuity result available for C's derived utility from the game is much weaker, though still strong enough for my purposes. Let Z(c,d) denote the set of solutions of (2.B), which is compact because U is continuous, define

$$(2.8) \qquad V^C(c,d) \equiv \max_{z \in Z(c,d)} U(c,\underline{1}-z),$$

call the problem in (2.6) "(2.C)", and let z(c,d) denote a solution of (2.C). If D breaks any ties for the solution of (2.B) by dividing as C would prefer, V^C is the function that gives C's derived utility from the game. We can now establish the following theorem:

Theorem 2.8: <u>When c=d, V^C is jointly continuous in c and d.</u>

Proof: Let $\{(c_i,d_i)\}$ be a sequence in I^{2p} that converges to (c_0,d_0), where $c_0=d_0$. Since $V^D(c_i,d_i)$ is everywhere continuous by Theorem 2.7, for every $\varepsilon > 0$ there exists a δ_1 such that, if (c_i,d_i) is in a δ_1 - neighborhood of (c_0,d_0),

$$(2.9) \quad |V^D(c_i,d_i) - V^D(c_0,d_0)| = |U(d_i,z(c_i,d_i)) - U(d_0,z(c_0,d_0))| \leq \varepsilon/2.$$

Further, since $\{(c_i,d_i)\} \to (c_0,d_0)$ and $c_0=d_0$, by the continuity of U there exists a δ_2 such that if (c_i,d_i) is in a δ_2 - neighborhood of (c_0,d_0),

$$(2.10) \qquad |U(c_i,z(c_i,d_i)) - U(d_i,z(c_i,d_i))| \leq \varepsilon/2.$$

Thus, if $\delta \leq \min\{\delta_1,\delta_2\}$, it follows from (2.9), (2.10), the Triangle Inequality, and Theorem 2.1 applied to (2.B) that

$$(2.11) \quad |V^C(c_i,d_i) - V^C(c_0,d_0)| = |U(c_i,\underline{1}-z(c_i,d_i)) - U(c_0,\underline{1}-z(c_0,d_0$$

$$|U(c_i,z(c_i,d_i)) - U(d_0,z(c_0,d_0))| \leq$$

$$|U(c_i,z(c_i,d_i)) - U(d_i,z(c_i,d_i))| +$$

$$|U(d_i,z(c_i,d_i)) - U(d_0,z(c_0,d_0))| \leq \varepsilon.$$

Thus, if $c_0=d_0$, for every $\varepsilon > 0$ a δ can be chosen such that if (c_i,d_i) is in a δ - neighborhood of (c_0,d_0), $V^C(c_i,d_i)$ is in an ε - neighborhood of $V^C(c_0,d_0)$, which completes the proof. $\square$

The intuition behind Theorem 2.8 is almost as simple as that behind Theorem 2.7. As with V^D, whenever D's optimal division varies continuously with preferences, V^C is continuous. By Theorem 2.7, when the optimal division changes discontinuously, D must be indifferent about the change. When C's preferences are identical to D's, C must therefore also be indifferent about the change, so when c=d, discontinuous changes in D's optimal division do not induce discontinuities in V^C. It is interesting to note that, while discontinuities in V^C cannot be ruled out in general, the smaller the difference between C's and D's preferences, the smaller are the possible discontinuities.

The analysis of this section has provided a partial resolution of the puzzle mentioned in the introduction, and has identified a class of cases where application of the divide and choose method can be strongly recommended. The conspicuous failure to use the method in pure-trade situations is probably rational in some cases. At least when preferences are convex, there exist allocations, notably the EICE, that are both more efficient and treat agents more equally than the allocations generated by the divide and choose method.[12] Casual empiricism suggests that when there are only two agents whose strategic possibilities are not limited by competition, EICE allocations are much more costly to achieve than divide and choose allocations, but their greater equity and efficiency may be worth the extra cost in some cases.

However, the divide and choose method does possess impor tant optimality properties -- fairness and efficiency in the fair set -- which make it an attractive alternative to other arbitratio methods. The non-Pareto-efficiency of the method is less disturbin than it might otherwise be because bargaining costs have been left out of the model, and the unequal treatment can be mitigated by assigning the role of divider randomly. Additional perspective can be gained by comparing the divide and choose method with the naive alternative of ED. Both appear to require roughly the same level of cooperation and administrative effort -- very little -- and both generate fair allocations that need not be Pareto-efficient. The method treats agents less equally than does ED, but the fact that it generates an allocation that is efficient in the set of fair allocations is an important advantage over ED, which is completely independent of agents' preferences and may therefore be extremely inefficient. On balance, the divide and choose methods seems superior to ED as a fair division device. Thus, in situations when bargaining costs are likely to be high and the inefficiency and unequal treatment of the divide and choose method are not likely to be great, its use can be recommended. These conditions obtain in the case of nearly identical preferences, where I have shown that the method's drawbacks almost completely disappear. Noncooperative agents with similar preferences playing the game will reach a settlement nearly as equitable and efficient as any feasible allocation, and there is no question that the method should be used in this case.

An important real-world problem for which the analysis presented here seems appropriate is that of divorce settlements. The assumptions of perfect knowledge and noncooperative behavior seem particularly palatable here. To the extent that assets like houses and cars can be readily liquidated, the assumption of perfect divisibility seems fairly reasonable as well. Children are, of course, divisible, although not in the Old Testament sense! One might even argue that married couples are likely to have similar preferences.

For obvious reasons, divorce settlements are often difficult to negotiate, and as a result, large sums are expended on them every year. Therefore, unless there is some reason to believe that exploitation by the divider will be a serious problem (as might happen, for example, if one parent had a very strong attachment to a child that was not shared by the other parent, in which case he or she might, as chooser, receive custody of the child for a little more than half the time and almost nothing else) a good case for experimenting with the divide and choose method as a technique for negotiating divorce settlements can be made.

3. THE DIVIDE AND CHOOSE GAME WITH PRODUCTION

Although the assumption that pure trade only is possible is a convenient simplification, most important economic situations involve production. In this section the analysis of Section 2 is extended to the case where production is possible. As in Section 2, agents are assumed to behave noncooperatively and to have perfect knowledge of all production and utility functions, and C is assumed to choose as D would prefer whenever he is indifferent between the possibilities offered him.

Each person is endowed with the vector $\underline{1}$ of perfectly divisible nonproduced goods that may either be consumed directly or applied in production. It may be helpful to think of these goods as different kinds of labor; one person's labor of a particular type may not be perfectly substitutable for the other's in production. There is also a vector of homogeneous and perfectly divisible produced goods, whose net outputs depend only on the inputs of each person's various kinds of labor. Each person's preferences are representable by a utility function that depends continuously on his labor supply vector, denoted by x^i, (i=C,D), and on his consumption of produced goods, denoted by y^i, (i=C,D). Since there is no danger of confusion with the notation of the previous section, let $U^D(\underline{1}-x^D,y^D)$ and $U^C(\underline{1}-x^C,y^C)$ denote D's and C's utility functions, which are both assumed increasing in all arguments. Let

$$(3.1) \qquad y = G[\lambda,x^D,x^C]$$

be the vector-valued function that gives the maximum possible outputs for given inputs and for a given value of the parameter vector $\lambda \equiv y/\underline{1}\cdot y$. Heuristically, this function can be determined by following the ray associated with a value of λ out to the production possibility frontier determined by x^D and x^C; varying λ sweeps out the function G. I assume that $G[\lambda,\underline{0},\underline{0}] = \underline{0}$ for all λ and that all inputs are productive. It follows that all components of G are strictly increasing in x^D and x^C, except when some component of λ equals zero, in which case the corresponding component of G is zero for all values of x^D and x^C. G gives <u>net</u> outputs -- if intermediate inputs are necessary, they are implicit in the form of the function, as are all the usual production efficiency conditions.

A division by D is represented by the vector $\omega \equiv (x_1,x_2,\lambda,\theta)$, where x_i is the vector of labor supplies required of the player who assumes role i, and θ is the vector of shares of total production received by the player who assumes role 1. λ and θ could be allowed to depend on which player assumes role 1, in effect allowing the divider to plan in terms of total production and consumption instead of proportions and shares. This generalization would leave all arguments and conclusions (though not the outcome of the game) essentially unchanged, but the present formulation is retained for its relative simplicity.

In this section, a <u>fair</u> allocation is defined as one where no agent would prefer to exchange roles (i.e., required labor supplies and consumption shares) with another.[13] Let $U^D(i)$ denote the level of utility obtained by D if he assumes role i, (i=1,2), and define $U^C(i)$ in the same way. If D has made the division ω, the reader may verify that:

$$(3.2)\qquad U^D(1) = U^D(\underline{1}-x_1, \theta G[\lambda, x_1, x_2]),$$

$$(3.3)\qquad U^D(2) = U^D(\underline{1}-x_2, (\underline{1}-\theta)G[\lambda, x_2, x_1]),$$

$$(3.4)\qquad U^C(1) = U^C(\underline{1}-x_1, \theta G[\lambda, x_2, x_1]), \text{ and}$$

$$(3.5)\qquad U^C(2) = U^C(\underline{1}-x_2, (\underline{1}-\theta)G[\lambda, x_1, x_2]),$$

recalling that $\theta G \equiv (\theta_1 G_1, \ldots, \theta_s G_s)$, where the θ_i and G_i are the components of θ and G. Assuming without loss of generality that D assumes role 1 at the solution, D's problem can be written as follows:

$$\text{(3.A)} \qquad \max_{\underline{0} \leq \omega \leq \underline{1},\, \lambda \cdot \underline{1}=1} U^D(1) \text{ s.t. } U^C(1) \leq U^C(2).$$

The following theorem is analogous to Theorem 2.1, but its proof is somewhat more involved.

Theorem 3.1: <u>Any solution to (3.A) must satisfy</u>

$$\text{(3.6)} \qquad U^C(1) = U^C(2).$$

Proof: The proof is by contradiction. Suppose that $\tilde{\omega} \equiv (\tilde{x}_1, \tilde{x}_2, \tilde{\lambda}, \tilde{\theta})$ is a solution to (3.A) at which $U^C(1) < U^C(2)$. I shall argue that there always exists a feasible ω in the neighborhood of $\tilde{\omega}$ at which $U^D(1)$ is higher than at $\tilde{\omega}$, contradicting the assumption that $\tilde{\omega}$ is a solution of (3.A). By the continuity of U and G, any sufficiently small variation of ω around $\tilde{\omega}$ still satisfies the constraint $U^C(1) \leq U^C(2)$, so in checking feasibility we need only insure that $\underline{0} \leq \omega \leq \underline{1}$ and $\lambda \cdot \underline{1} = 1$. The reader may verify from (3.2) that it is possible to increase $U^D(1)$ by raising some component of x_2 above $\tilde{x}_2$ unless either

(i) $\tilde{x}_2 = \underline{1}$ or (ii) $\tilde{\lambda}\tilde{\theta} = \underline{0}$.

If $\tilde{x}_1 = \tilde{x}_2 = \underline{0}$, substitution into (3.4) and (3.5) yields $U^C(1) = U^C(2)$ for $\omega = \tilde{\omega}$, and there is nothing more to prove. Otherwise, since all inputs are productive and $G[\tilde{\lambda},\underline{0},\underline{0}] = \underline{0}$, $G_i[\tilde{\lambda},\tilde{x}_1,\tilde{x}_2] > 0$ for all i such that $\tilde{\lambda}_i > 0$. Thus, $U^D(1)$ may be increased while preserving feasibility by raising some component of θ above $\tilde{\theta}$ unless $\tilde{\lambda}(\underline{1}-\tilde{\theta}) = \underline{0}$. But $\tilde{x}_2 = \underline{1}$ in case (i), so if $\tilde{\lambda}(1-\tilde{\theta}) = \underline{0}$, (3.5) implies that $U^C(2) = U^C(\underline{0},\underline{0})$, contradicting the assumption that $U^C(1) < U^C(2)$ for $\tilde{\omega}$ satisfying $\underline{0} \leq \tilde{\omega} \leq 1$ and strongly monotonic U^C. Finally, in case (ii) $\tilde{\lambda}\tilde{\theta} = \underline{0}$, so $\tilde{\lambda}(\underline{1}-\tilde{\theta}) = \tilde{\lambda}-\tilde{\lambda}\tilde{\theta} = \underline{0}$ cannot hold unless $\tilde{\lambda} = \underline{0}$, which is impossible since $\tilde{\lambda}\cdot\underline{1} = 1$. This completes the proof. □

As in the pure-trade case, the outcome of the game is not necessarily a Pareto-efficient allocation. The proofs of the following lemma and theorems are identical to their counterparts in Section 2, and so are omitted.

Lemma 3.1: <u>By dividing appropriately, D can enforce any feasible fair allocation as the outcome of the divide and choose game with production.</u>

Theorem 3.2: <u>The outcome of the divide and choose game with production is a fair allocation, and if D agrees to</u>

break ties for the solution of (3.A) by dividing as C would prefer, the outcome is an efficient point in the set of all fair allocations.

Theorem 3.3: In the divide and choose game with production D always does at least as well in the role of divider as he would in the role of chooser.

In a productive economy, an EICE can be defined as a competitive equilibrium that results when each agent's initial endowment consists of all of each of his nonproduced goods plus his share of society's initial stock of produced goods. (The latter equals 0 in the present model.) At this point in the discussion, it would be natural to compare the allocations obtained from the divide and choose game with production with EICE allocations, and with the ED allocation, as was done in the pure-trade case. However, when there is production, no single definition of ED seems compelling, and even if an EICE exists the analog of Theorem 2.4 is not true. The problem is that an EICE defined in this way is not necessarily fair if agents have different productivities.[14] If C envies D at an EICE allocation, D cannot enforce it as the outcome of the game, and examples can easily be constructed to show that D may prefer his EICE allocation to his game allocation. Similarly, the second half of the proof of Theorem 2.4 fails to go through here, so C may prefer his game allocation to his EICE allocation. Thus, in a productive economy, it is possible for the divide and choose method to treat agents more equally than the EICE.

The remaining results in Section 2 also generalize to the case of production. They are recorded here without their proofs, which parallel those of the corresponding results in Section 2.

Theorem 3.4: <u>If D and C have identical preferences and productivities, they are treated equally in the divide and choose game with production</u>.

Theorem 3.5: <u>When D and C have identical preferences and productivities, the outcome of the divide and choose game with production is a Pareto-efficient allocation</u>.

The requirement that D and C have identical productivities means that there must be perfect substitutability in production between their labors of each type.

For the next two theorems, the functions V^D and V^C defined in Section 2 should be reinterpreted in the obvious way, and the parameter vectors c and d taken to include components that parameterize preferences (including preferences about labor supplies) and productivities, in such a way that, as the components of c and d that parameterize the production functions approach each other, D's and C's labors of each type approach perfect substitutability.

Theorem 3.6: When production is possible, V^D is everywhere jointly continuous in c and d.

Theorem 3.7: When production is possible and c=d, V^C is jointly continuous in c and d.

Theorems 3.6 through 3.7 have the same interpretation as their counterparts in the pure-trade case: when agents have similar preferences and productivities, the divide and choose method treats them similarly and generates an allocation that is nearly Pareto-efficient.

The introduction of production radically alters the relationship between the divide and choose method and the fair division devices it was compared with in the pure-trade case. ED appears to lose its precise meaning with production, as no single way of resolving conflicts of interest about production levels seems compelling. In a pure-trade economy, at least if preferences are convex, the goals of equity and efficiency are compatible and could be satisfied by the EICE if a costless process were available to locate it. The pure-trade EICE shares the divide and choose method's optimality properties of fairness and efficiency in the fair set, and in addition treats agents equally and generates Pareto-efficient allocations, advantages not shared by the method unless preferences are identical. But when production is possible, an EICE need not be fair unless agents have identical productivities; in fact, simultaneously fair and Pareto-efficient allocations may fail to exist even in very regular economies.[15] EICE allocations are still Pareto-efficient, of course, but with production they may treat agents less equally than the divide and choose method's allocations. In contrast, with production the divide and choose method retains all of its optimality properties. Efficiency in the set of fair allocations is a very desirable property if equity is considered as important as efficiency in judging allocations, and may be the best that can be hoped for in general since fairness and full Pareto-efficiency are not always compatible in a productive economy. Of course, the divide and choose method still favors the divider at the expense of the chooser, but this asymmetry is no longer a disadvantage relative to the EICE, which now favors agents with high productivities. The EICE treats agents unequally unless they have identical productivities, and the divide and choose method treats them unequally unless they have identical preferences <u>and</u> productivities. And it is still possible to reduce the inequality of the divide and choose method by assigning roles randomly.

To sum up, while the divide and choose method retains all of its desirable properties when production is added to the model, alternative fair division devices lose many of theirs. If anything, this attrition makes the divide and choose method more attractive when production is possible than in a pure-trade economy.

4. THE DIVIDE AND CHOOSE GAME WITH COOPERATIVE TRADE

Because the allocation resulting from the divide and choose game is not necessarily Pareto-efficient, mutually beneficial trades may take place after the divide and choose process. C may also attempt to influence D's division by threatening to choose the "wrong" bundle. The analysis of Section 2 does not presuppose that such phenomena do not occur, but it does implicitly assume that players do not anticipate them in choosing their strategies. Since this may not always be a reasonable assumption, this section will discuss the implications of optimizing behavior in the pure-trade case when players expect cooperative trade of a particular kind to follow the divide and choose phase of the game.

Unless otherwise noted, the assumptions and notation of Section 2 are continued in this section. The indeterminacy that pervades cooperative game theory will be resolved by applying a predetermined rule like that suggested by Nash [7] for such games. Players play the divide and choose game to establish their threats $\bar{U}^D$ and $\bar{U}^C$ -- the utilities of the bundles they receive if no trade follows the divide and choose process. Then bargaining takes place, and the utilities resulting from the agreement ultimately reached are assumed to be continuous functions of $\bar{U}^D$ and $\bar{U}^C$, denoted by $W^D[\bar{U}^D,\bar{U}^C]$ and $W^C[\bar{U}^D,\bar{U}^C]$. W^D is strictly increasing in $\bar{U}^D$ and strictly decreasing in $\bar{U}^C$, and W^C is strictly decreasing in $\bar{U}^D$ and strictly increasing in $\bar{U}^C$. It is probably reasonable to assume in addition that W^D and W^C single out an allocation in the core, but this is unnecessary for the results of this section. The present assumption about the bargaining process includes as a special case the Nash cooperative solution, which has several desirable properties (described in Nash [7]) and seems to be a sensible solution concept for two-person cooperative games. Nash assumed that players' utility functions are cardinal in the von Neumann-Morgenstern sense, but that assumption is not needed here. Of course, the forms of W^D and W^C depend, like those of social welfare functions, on the particular ordinal indicator of utility chosen.

Again assuming that z is the bundle D intends for himself in the first phase of the game, D's optimal strategy is determined by solving the following programming problem:

$$\text{(4.A)} \qquad \max_{\underline{0} \leq z \leq \underline{1}} W^D[U^D(z), U^C(\underline{1}-z)]$$

$$\text{s.t.} \quad W^C[U^D(\underline{1}-z), U^C(z)] \leq W^C[U^D(z), U^C(\underline{1}-z)].$$

Because the objective function of (4.A) is monotone increasing in z, an argument like that of Section 2 establishes the following theorem:

Theorem 4.1: <u>Any solution to (4.A) must occur when</u>

$$\text{(4.1)} \qquad W^C[U^D(\underline{1}-z), U^C(z)] = W^C[U^D(z), U^C(\underline{1}-z)].$$

W^D and W^C, viewed as functions of z, satisfy the same monotonicity and continuity conditions placed on U^D and U^C in the analysis of Section 2, so (4.A) is formally analogous to (2.A). Therefore, the analogs of Lemma 2.1 and Theorem 2.2 could be established here and used to show that the role of divider is an advantage in this game. However, I will take this opportunity to illustrate a direct proof, due to Martin Weitzman, which is equally general and less unnatural in this context.

Theorem 4.2: <u>In the divide and choose game with cooperative trade, D always does at least as well in the role of divider as he would in the role of chooser.</u>

Proof: Consider (4.A) and the analogous problem C would face if he were divider:

$$\text{(4.B)} \qquad \max_{0 \leq z \leq 1} W^C[U^D(z), U^C(\underline{1}-z)]$$

$$\text{s.t.} \quad W^D[U^D(\underline{1}-z), U^C(z)] \leq W^D[U^D(z), U^C(\underline{1}-z)].$$

Let z^* denote a solution to (4.A), and $\hat{z}$ denote a solution to (4.B). The proof that $W^D[U^D(z^*), U^C(\underline{1}-z^*)] \geq W^D[U^D(\hat{z}), U^C(\underline{1}-\hat{z})]$ is by contradiction. Assume the contrary. Then $\hat{z}$ must not be feasible for (4.A):

$$W^C[U^D(\underline{1}-\hat{z}),U^C(\hat{z})] > W^C[U^D(\hat{z}),U^C(\underline{1}-\hat{z})]. \tag{4.2}$$

(4.2) implies in turn that $\underline{1}-\hat{z}$ is not feasible for (4.B):

$$W^D[U^D(\hat{z}),U^C(\underline{1}-\hat{z})] > W^D[U^D(\underline{1}-\hat{z}),U^C(\hat{z})]. \tag{4.3}$$

But Theorem 4.1 applied to (4.B) implies that

$$W^D[U^D(\hat{z}),U^C(\underline{1}-\hat{z})] = W^D[U^D(\underline{1}-\hat{z}),U^C(\hat{z})], \tag{4.4}$$

which contradicts (4.3). □

While D's optimal strategy in the divide and choose game with cooperative trade is usually quite different from his optimal strategy in the simple divide and choose game, these strategies share some properties, since the problems he must solve to find them are formally analogous. However, whether or not the outcome of the game studied in this section is a fair allocation, a Pareto-efficient allocation, or has any particular relationship to EICE or ED allocations depends on the nature of the bargaining process underlying the functions W^D and W^C.

5. FURTHER GENERALIZATIONS; CONCLUSIONS

Several possible further generalizations would be of interest. Combining the analyses of production and cooperative trade seems unlikely to present any new difficulties. Allowing some indivisible goods would be an important step toward practical applications, and is probably possible.[16] Assuming that D does not know C's preferences with certainty, but has uncertain prior beliefs about them (perhaps because he knows that C is drawn from a population for which he knows only the distribution of preferences) would be of great interest. It is plausible that if D knows that C's preferences are likely to differ systematically from his own he can exploit this knowledge to obtain an ex ante advantage. Also, since D must now weigh the advantages of exploitation against the chance that he will go too far and accidentally receive the "wrong" bundle, it is likely that increasing uncertainty reduces his advantage in the game. However, it seems very difficult to formalize and prove these notions without placing arbitrary restrictions on D's prior distribution about C's preferences.

There are at least two ways that the pure-trade divide and choose procedure might be generalized to the n-person case. Following Steinhaus [10], we could study a process in which the divider cuts a "slice," and each chooser, in a prespecified order, has the opportunity to diminish the slice if he wishes. The last person to diminish a slice must take it as his share, and the game then begins anew without him. An administratively simpler and analytically more tractable procedure would be to allow the divider to divide the bundle into n parts from which the other players choose in a prespecified order, the divider receiving as his share the part that remains after all the others have chosen. My discussion will be confined to this process.

When the problem that determines the divider's optimal division in the n-person case is formulated, it takes the form

$$(5.A) \qquad \underset{\sum_{i=1}^{n} z^i = \underline{1}}{\text{Max}} \quad U^n(z^n) \qquad \text{s.t.} \quad U^k(z^j) \leq U^k(z^k), \quad (k=1,\ldots,n-1; j=k+1,\ldots,n),$$

where the ith player has utility function U^i, and receives the bundle z^i, $(i=1,\ldots,n)$. Player n is the divider, and the other n-1 players choose in the order of their indices, starting with player 1. The inequality constraints of (5.A) express the requirement that each chooser must freely choose "his" bundle from those available at his turn to choose, subject only to the restriction that he breaks ties as n directs.

It is clearly still true that the divider can enforce any feasible fair allocation in the n-person divide and choose game, and if it were possible to show that solutions of (5.A) necessarily involved all of the inequality constraints holding with equality, an argument could be constructed to show that the n-person game always generates a fair allocation. Given these two results, virtually all of the results of Section 2 could easily be generalized to the n-person case. Unfortunately, only a weaker generalization of Theorem 2.1 is possible.[17] For example, in the three-person case it is only possible to show, under assumptions like those of Section 2, that one of the inequality constraints (k=2;j=3) of (5.A) is necessarily binding at the solution, and that at least one of the other two is also binding. I have constructed a three-person, two-good counterexample to show that even with smooth, convex and strongly monotonic preferences it is possible to have solutions of (5.A) where one constraint is slack, that the outcome of the game need not be a fair allocation, and that the role of divider need not be an advantage. As often happens in economic theory, the regularity of the two-person case does not extend to the n-person case. As a result, the divide and choose method seems to have much less to recommend it when there are more than two persons.

In this essay I have shown the divide and choose method to have several interesting properties. The noncooperative equilibrium of the game induced by the method in the pure-trade case need not be a Pareto-efficient allocation, but it is a fair allocation and an efficient point in the set of all fair allocations. The game confers an advantage on the divider, and its allocations treat agents less equally than EICE and ED allocations. When both agents have identical preferences, however, the Pareto-inefficiency, divider's advantage, and unequal treatment all disappear, and these results are robust in the sense that, if agents have nearly identical preferences, nearly equal treatment and near-Pareto-efficiency result.

Nearly all of the pure-trade results can be extended to the case where production is possible, and attrition of the optimality properties of alternative arbitration devices seems to make the divide and choose method even more attractive in the production case than in the pure-trade case. When the divide and choose process is followed by cooperative trade that players anticipate in their strategy choices, some of the results, notably the existence of divider's advantage, go through. Finally, virtually none of the strong results obtained in the two-person case can be generalized to n-person versions of the divide and choose method.

Thus, the theoretical evidence tends to favor wider application of the divide and choose method to resolve two-person bargaining disputes. In the pure-trade case the method is fairly attractive in general, and extremely attractive when agents are in a nearly symmetrical position, since in this case it achieves, at very low administrative cost, an allocation that is nearly as equitable and efficient as any feasible allocation. While failure to use the divide and choose method is probably rational in some cases due to its Pareto-inefficiency and unequal treatment, at least one class of cases has been identified where it can definitely be recommended: when preferences are similar the difference in administrative costs between the divide and choose method and alternative procedures is almost certain to outweigh its disadvantages. In Section 2 I argued that divorce settlements might fall into this class of cases frequently. To the extent that they do, large increases in welfare could be achieved by using the divide and choose method.

FOOTNOTES

[1]See, for example, [3, 5, 6, pp. 363-369; 8, p. 85; 9, and 10]. It may be possible to argue that Rawls [8, pp. 16, 18, 103, 136-137, 152-153] has a divide and choose procedure in mind when he argues that rational men in the "original position" would accept his principles of justice. However, he also makes it clear that the "veil of ignorance" covers even one's own preferences, so the analysis of this paper, which assumes preferences are known, does not apply.

[2]This sense of the word "fair" is consistent with the usage in [3, 4, 5, and 10]. In [12], our "fair" is equivalent to "equitable", and "fair" means simultaneously "equitable" and Pareto-efficient.

[3]See, for example, [3, 5, and 10], which are mostly concerned with generalizing the result to the n-person case.

[4]In a pure-trade economy, an EICE is defined as a competitive equilibrium at which all agents' initial endowments have equal value. For a discussion of the EICE as a fair division device, see [12].

[5]Martin Weitzman has pointed out to me that this assumption can be greatly weakened. For many of the results it is clear from the

proofs that an assumption of local nonsatiation relative to the set of feasible allocations will suffice. The assumption of perfect divisibility can probably also be relaxed somewhat. I have ignored these possibilities for the sake of clarity.

[6] This result does not really depend on C's cooperation when he is indifferent. If D is indifferent between the bundles of his division, it does not matter whether C cooperates. If D strictly prefers one of the bundles and induces the desired choice by making a small adjustment in his division, it is clear from the continuity of U^D that the resulting allocation will still be fair.

[7] Martin Weitzman has provided an alternative proof of this theorem, which is illustrated in Section 4.

[8] The first part of this theorem was originally conjectured by Clifford Donn.

[9] This result is part of Theorem 2.2 in [12, p. 68].

[10] The assumption that preferences can be represented by utility functions depending on only a finite number of parameters is more restrictive than the assumptions needed in the rest of Section 2, but relaxing it does not seem worth the technical effort required. The assumption that the same number of parameters

is needed to describe both utility functions is inessential, since some parameters can always be held constant.

[11] See[1, pp. 115-116] or [2, p. 19]. Berge's Maximum Theorem is paraphrased here for convenience: Let $\alpha \varepsilon A$ be a vector of parameters, let $\beta \varepsilon B$ be a vector of control variables, and let f be a continuous real-valued function defined on $A \times B$. Define

$$M(\alpha) \equiv \underset{\beta \varepsilon h(\alpha)}{\text{Max}}\ f(\alpha,\beta),$$

where $h(\alpha)$ is a closed nonempty subset of B for each $\alpha \varepsilon A$, and h is continuous for each $\alpha \varepsilon A$. Then M is a continuous function of α.

In problem (2.B), the upper hemi-continuity of the feasible-region correspondence follows immediately; its lower hemi-continuity follows, strictly speaking, from results in Heller [4a]. Even though the feasible region is not generally convex; it does satisfy the weaker condition of "local interiority" defined and shown to suffice there.

[12] For a discussion of the existence of simultaneously fair and Pareto-efficient allocations in a pure-trade economy, see [12, pp. 68-70], and recall the difference in terminology pointed out in n.2.

[13] In [12, p. 71], Varian defines fairness in an economy with production in terms of total consumption. If I allowed λ and θ to depend on which player assumes role 1, the present model would differ from Varian's only in notation, and none of the resu would change.

[14] This problem still arises if we instead define an EICE as a competitive equilibrium where all agents' initial endowments have equal values, because agents' leisures will have different prices if there are productivity differences. An agent may envy another for his high leisure consumption, but be unwilling to buy as much for himself if his high productivity makes his own leisure more expensive. It is possible to weaken the concept of fairness so that an EICE as defined in this paper is "fair," although not in a completely satisfactory way, and not for genera production technologies. For a discussion of the problems encountered in defining fairness and the EICE in an economy with production, see [7a] and [12, pp. 70-75] and recall the differenc in terminology pointed out in n.2.

[15]See [7a] and [12, pp. 71-72].

[16]See [5, pp. 33-35] and [6, pp. 366-367] for a discussion of methods for handling indivisibilities.

[17]I am indebted to an anonymous referee for pointing out the fatal error in my "proof" of the strong n-person generalization of Theorem 2.1.

REFERENCES

[1] Berge, C.: Topological Spaces. New York: MacMillan, 1963.

[2] Debreu, G.: Theory of Value. New York: John Wiley and Sons, Inc., 1959.

[3] Dubins, L. E., and E. H. Spanier : "How to Cut a Cake Fairly," American Mathematical Monthly, 68 (1961), 1-17.

[4] Feldman, A., and A. Kirman : "Fairness and Envy," American Economic Review, 64 (1974), 995-1005.

[4a] Heller, W. P. : "Continuity in General Nonconvex Economies (with Applications to the Convex Case)," in G. Schwödiauer (ed.), Equilibrium and Disequilibrium in Economic Theory. Dordrecht, Holland: D. Reidel Publishing Company, 1977.

[5] Kuhn, H. W. : "On Games of Fair Division," in M. Shubik (ed.), Essays in Mathematical Economics in Honor of Oskar Morgenstern. Princeton: Princeton University Press, 1967.

[6] Luce, R. D., and H. Raiffa: Games and Decisions: Introduction and Critical Survey. New York: John Wiley and Sons, Inc., 1957.

[7] Nash, J.: "Two-Person Cooperative Games," Econometrica, 21(1953), 128-140.

[7a] Pazner, E. A., and D. Schmeidler: "A Difficulty in the Concept of Fairness," Review of Economic Studies, 41 (1974), 441-443.

[8] Rawls, J.: A Theory of Justice. Cambridge: Harvard University Press, 1971.

[9] Singer, E.: "Extension of the Classical Rule of Divide and Choose," Southern Economic Journal, 28(1962), 391-394.

[10] Steinhaus, H.: "The Problem of Fair Division," Econometrica 16 (1948), 101-104.

[11] Twain, M.: Life on the Mississippi. New York: Washington Square Press, 1968.

[12] Varian, H. R.: "Equity, Envy, and Efficiency," Journal of Economic Theory, 9 (1974), 63-91.

For Product Safety Concerns and Information please contact our EU representative GPSR@taylorandfrancis.com Taylor & Francis Verlag GmbH, Kaufingerstraße 24, 80331 München, Germany